Pre -Algebra Practice
1st Edition

equals(me) packets give exciting and challenging entertainment while improving ones mind skills.

equals(me) packets are designed to assist in learning math and algebra concepts. These packets provide excellent algebra practice.

Each packet includes a variety of math and algebra techniques. The packets are modeled using the Study-Try-Compare methods. This gives you the opportunity to study the steps in solving, try for yourself, and compare results. You can also try to solve and then compare your solution.

For school, college, work, or at home, equals(me) packets make math fun.

ISBN: 978-1983752308

0418rev3.1

Combining like terms

Group all like terms together.
Keep (+) or (-) signs with number.

$$3x - 2 + 6x + 5 - 2x$$
$$(+3x) \ (-2) \ (+6x) \ (+5,) \ (-2x)$$

Add or subtract like terms.

$$3x + 6x - 2x - 2 + 5$$
$$(3x + 6x - 2x) - 2 + 5$$
$$7x + 3$$

Prime Factorization

Find 2 numbers in which their product equals your number.

Continue to factor until all numbers are prime.

$$60$$
$$2 \times 30$$
$$2 \times 2 \times 15$$
$$2 \times 2 \times 3 \times 5$$

Write as exponential notation.

$$2^2 \times 3 \times 5$$

Solve equations

Get variable to left side and everything else to right side.

$$3x + 5 = 2x + 8$$
$$3x - 2x = 8 - 5$$
$$1x = 3$$
$$x = 3$$

If they are added to the one side, subtract on the other side.
If they are subtracted to the one side, add to the other side.
If they are multiplied to the one side, divide on the other side.
If they are divided on the one side, multiply to the other side.

Scientific Notation

$1.23 \times 10^3 = 1230$ $1.23 \times 10^{-1} = 0.123$

$1.23 \times 10^2 = 123$ $1.23 \times 10^{-2} = 0.0123$

$1.23 \times 10^1 = 12.3$ $1.23 \times 10^{-3} = 0.00123$

When multiplying, add exponents.

$$\begin{array}{r} 1.3 \times 10^5 \\ \underline{\times\ 1.2 \times 10^2} \\ 1.56 \times 10^7 \end{array}$$

When adding/subtracting, manipulate numbers to have the same exponent.

$$\begin{array}{r} 1.3 \times 10^5 \\ \underline{+\ 1.2 \times 10^4} \end{array}$$

If exponent is raised, decimal is moved left.
If exponent is lowered, decimal is moved right.

$$\begin{array}{r} 1.30 \times 10^5 \\ \underline{+\ .12 \times 10^5} \end{array}$$

PEMDAS

Solve in PEMDAS order.

1. *Parenthesis*
2. *Exponents*
3. *Multiply/Divide (left to right)*
4. *Add/Subtract (left to right)*

$(3 + 2) \times 2^2 + 4$

$(5) \times 2^2 + 4$

$(5)(4) + 4$

$20 + 4$

24

See pages 78 & 79 for more rules of algebra.

Prime Factorization

100

2 x 50

2 x 2 x 25

2 x 2 x 5 x 5

$$2^2 \; x \; 5^2$$

80

2 x 40

2 x 2 x 20

2 x 2 x 2 x 10

2 x 2 x 2 x 2 x 5

$$2^4 \; x \; 5$$

140

2 x 70

2 x 2 x 35

2 x 2 x 5 x 7

$$2^2 \; x \; 5 \; x \; 7$$

36

6 x 6

2 x 3 x 2 x 3

$$2^2 \; x \; 3^3$$

Prime Factorization

100

_____

_____

_____

_____

_____

80

_____

_____

_____

_____

_____

140

_____

_____

_____

_____

_____

36

_____

_____

_____

_____

_____

Prime Factorization

144

4 x 36

2 x 2 x 6 x 6

2 x 2 x 2 x 3 x 2 x 3

$$2^4 \, x \, 3^2$$

180

2 x 90

2 x 2 x 45

2 x 2 x 5 x 9

2 x 2 x 5 x 3 x 3

$$2^2 \, x \, 3^2 \, x \, 5$$

130

2 x 65

2 x 5 x 13

168

4 x 42

2 x 2 x 6 x 7

2 x 2 x 2 x 3 x 7

$$2^3 \, x \, 3 \, x \, 7$$

Prime Factorization

144

180

130

168

Prime Factorization

220

11 x 20

11 x 2 x 10

11 x 2 x 2 x 5

2^2 x 5 x 11

240

6 x 40

2 x 3 x 4 x 10

2 x 3 x 2 x 2 x 2 x 5

2^4 x 3 x 5

260

13 x 20

13 x 5 x 4

13 x 5 x 2 x 2

2^2 x 5 x 13

245

5 x 49

5 x 7 x 7

7^2 x 5

Prime Factorization

220

240

260

245

Prime Factorization

162

2 x 81

2 x 9 x 9

2 x 3 x 3 x 3 x 3

3^4 x 2

290

5 x 58

5 x 2 x 29

64

8 x 8

2 x 4 x 2 x 4

2 x 2 x 2 x 2 x 2 x 2

2^6

320

4 x 80

2 x 2 x 8 x 10

2 x 2 x 2 x 4 x 2 x 5

2 x 2 x 2 x 2 x 2 x 2 x 5

2^6 x 5

Prime Factorization

162

_____

_____

_____

_____

_____

290

_____

_____

_____

_____

_____

64

_____

_____

_____

_____

_____

320

_____

_____

_____

_____

_____

PEMDAS - Simplify Expressions

$7 \times 3 - (14 - 2)$

$7 \times 3 - 12$

$21 - 12$

9

$\frac{14}{2} \times 5 + (6 \times 4)$

$\frac{14}{2} \times 5 + 24$

$7 \times 5 + 24$

$35 + 24$

59

$(\frac{36}{9}) \times (\frac{40}{8}) + 5$

$4 \times 5 + 5$

$20 + 5$

25

$8 \times 3 - (8 + 3)$

$8 \times 3 - 11$

$24 - 11$

13

PEMDAS - Simplify Expressions

$$7 \times 3 - (14 - 2)$$

$$\frac{14}{2} \times 5 + (6 \times 4)$$

$$\left(\frac{36}{9} \right) \times \left(\frac{40}{8} \right) + 5$$

$$8 \times 3 - (8 + 3)$$

PEMDAS - Simplify Expressions

$\frac{1}{2}$ x $\frac{1}{2}$ + 4 x 3 - 5

```
0.5 x 0.5 + 4 x 3 - 5
0.25 + 12 - 5
12.25 - 5
7.25
```

$(2 x 3)^2 + (\frac{3}{4} x 2)$

```
(6)² + (0.75 x 2)
(6)² + 1.5
 36 + 1.5
 37.5
```

$(2 x 8)^2 \div 4$

```
(16)²  ÷  4
 256 ÷ 4
 64
```

---

7 x (22 ÷ 2 x 4) - 3

```
7 x (11 x 4) - 3
7 x 44 - 3
308 - 3
305
```

PEMDAS - Simplify Expressions

$$\frac{1}{2} \times \frac{1}{2} + 4 \times 3 - 5$$

_____
_____
_____
_____
_____

$$(2 \times 3)^2 + (\frac{3}{4} \times 2)$$

_____
_____
_____
_____

$$(2 \times 8)^2 \div 4$$

_____
_____
_____
_____

$$7 \times (22 \div 2 \times 4) - 3$$

_____
_____
_____
_____

PEMDAS - Simplify Expressions

$(5 + 3) \times 5 \div 10 - 4$

$8 \times 5 \div 10 - 4$

$40 \div 10 - 4$

$4 - 4$

0

$8 \times (4 - 8) + \frac{1}{2}$

$8 \times (-4) + \frac{1}{2}$

$-32 + \frac{1}{2}$

$-32 + 0.5$

-32.5

$25 \div 5 + 10 - 2 + 5$

$5 + 10 - 2 + 5$

$15 - 2 + 5$

$13 + 5$

18

$(99 \div 11) \times (12 - 8) + 32$

$9 \times 4 + 32$

$36 + 32$

68

PEMDAS - Simplify Expressions

$$(5 + 3) \times 5 \div 10 - 4$$

_____
_____
_____
_____
_____

$$8 \times (4 - 8) + \tfrac{1}{2}$$

_____
_____
_____
_____
_____

$$25 \div 5 + 10 - 2 + 5$$

_____
_____
_____
_____
_____

$$(99 \div 11) \times (12 - 8) + 32$$

_____
_____
_____
_____
_____

PEMDAS - Simplify Expressions

$$24 - 6 \times 4 - \frac{21}{3}$$
$$24 - 24 - \frac{21}{3}$$
$$24 - 24 - 7$$
$$-7$$

$$30 - (7 + 5) \times 4 \div 2$$
$$30 - 12 \times 4 \div 2$$
$$30 - 48 \div 2$$
$$30 - 24$$
$$6$$

$$15 \times 3 \div \frac{25}{5}$$
$$45 \div \frac{25}{5}$$
$$45 \div 5$$
$$9$$

$$10^2 \div (5^2 \times 2)$$
$$10^2 \div (25 \times 2)$$
$$10^2 \div 50$$
$$100 \div 50$$
$$2$$

PEMDAS - Simplify Expressions

$$24 - 6 \times 4 - \frac{21}{3}$$

$$30 - (7 + 5) \times 4 \div 2$$

$$15 \times 3 \div \frac{25}{5}$$

$$10^2 \div (5^2 \times 2)$$

PEMDAS - Simplify Expressions

$45 \times 2 + 3^2 - 10$

$45 \times 2 + 9 - 10$

$90 + 9 - 10$

$99 - 10$

89

$(\frac{1}{2}) \times 40 \div 5$

$0.5 \times 40 \div 5$

$20 \div 5$

4

$3^3 - (\frac{25}{5}) \times 2$

$3^3 - 5 \times 2$

$27 - 5 \times 2$

$27 - 10$

17

$(55 \div 11) \times 8 - 2^3$

$5 \times 8 - 2^3$

$5 \times 8 - 8$

$40 - 8$

32

PEMDAS - Simplify Expressions

$$45 \times 2 + 3^2 - 10$$

$$(\tfrac{1}{2}) \times 40 \div 5$$

$$3^3 - (\tfrac{25}{5}) \times 2$$

$$(55 \div 11) \times 8 - 2^3$$

PEMDAS - Simplify Expressions

$$4 \times \left(\tfrac{3}{4}\right) - (5 - 3)^2$$

$$4 \times 0.75 - 2^2$$
$$4 \times 0.75 - 4$$
$$3 - 4$$
$$-1$$

$$1 + \tfrac{1}{2} \times 10 - 3^3$$

$$1 + \tfrac{1}{2} \times 10 - 27$$
$$1 + 5 - 27$$
$$6 - 27$$
$$-21$$

$$\left(8 \div \tfrac{1}{2}\right) \times \sqrt{4} - 2$$

$$16 \times 2 - 2$$
$$32 - 2$$
$$30$$

$$(5^2 - 5) + 2^2$$

$$(25 - 5) + 2^2$$
$$20 + 2^2$$
$$20 + 4$$
$$24$$

PEMDAS - Simplify Expressions

$$4 \times \left(\tfrac{3}{4}\right) - (5 - 3)^2$$

$$1 + \tfrac{1}{2} \times 10 - 3^3$$

$$\left(8 \div \tfrac{1}{2}\right) \times \sqrt{4} - 2$$

$$(5^2 - 5) + 2^2$$

PEMDAS - Simplify Expressions

$$\sqrt{9} \times 2^3 + 10$$

$3 \times 2^3 + 10$

$3 \times 8 + 10$

$24 + 10$

34

$$3^3 \times (\tfrac{15}{5} + 7)$$

$3^3 \times (3 + 7)$

$3^3 \times 10$

27×10

270

$$\sqrt{25} \times \sqrt{5^2} - 5$$

$\sqrt{25} \times \sqrt{25} - 5$

$5 \times 5 - 5$

$25 - 5$

20

$$(3^3 - 7) \times (2^2 + 1)$$

$(27 - 7) \times (4 + 1)$

20×5

100

PEMDAS - Simplify Expressions

$$\sqrt{9} \times 2^3 + 10$$

$$3^3 \times \left(\frac{15}{5} + 7 \right)$$

$$\sqrt{25} \times \sqrt{5^2} - 5$$

$$(3^3 - 7) \times (2^2 + 1)$$

PEMDAS - Simplify Expressions

$(13 + 5) \times (2 + 3 \times 2^2)$
$(13 + 5) \times (2 + 3 \times 4)$
$(13 + 5) \times (2 + 12)$
18×14
252

$80 \div 4 \times (2 + 3)$
$80 \div 4 \times 5$
20×5
100

$6 \times 6 \div 2 - 4 + 8$
$36 \div 2 - 4 + 8$
$18 - 4 + 8$
$14 + 8$
22

$(\frac{3 \times 3}{3}) \times 4 + 10$
$(\frac{9}{3}) \times 4 + 10$
$3 \times 4 + 10$
$12 + 10$
22

PEMDAS - Simplify Expressions

$$(13 + 5) \times (2 + 3 \times 2^2)$$

$$80 \div 4 \times (2 + 3)$$

$$6 \times 6 \div 2 - 4 + 8$$

$$\left(\frac{3 \times 3}{3} \right) \times 4 + 10$$

Combining Like Terms

$$2x - 9 + 5x + 10$$

$$2x + 5x - 9 + 10$$

$$7x - 9 + 10$$

$$7x + 1$$

$$4x + 2 - 6x + 8 - 1$$

$$4x - 6x + 2 + 8 - 1$$

$$-2x + 2 + 8 - 1$$

$$-2x + 9$$

$$7x - 5 - 1 - 6x + 8$$

$$7x - 6x - 5 - 1 + 8$$

$$1x - 5 - 1 + 8$$

$$1x + 2$$

$$x + 2$$

$$-12 + 2x + 3 - 5x$$

$$2x - 5x - 12 + 3$$

$$-3x - 12 + 3$$

$$-3x - 9$$

Combining Like Terms

$$2x - 9 + 5x + 10$$

_____

_____

_____

_____

_____

$$4x + 2 - 6x + 8 - 1$$

_____

_____

_____

_____

_____

$$7x - 5 - 1 - 6x + 8$$

_____

_____

_____

_____

_____

$$-12 + 2x + 3 - 5x$$

_____

_____

_____

_____

_____

Combining Like Terms

-4 + 9x - 5 - 8x + 1

$$9x \; - \; 8x \; - \; 4 \; - \; 5 \; + \; 1$$
$$1x \; - \; 4 \; - \; 5 \; + \; 1$$
$$1x \; - \; 8$$
$$x \; - \; 8$$

---

5 - 7 + 3x - 4x - 2

$$3x \; - \; 4x \; + \; 5 \; - \; 7 \; - \; 2$$
$$-1x \; + \; 5 \; - \; 7 \; - \; 2$$
$$-x \; - \; 4$$

4x - $2x^2$ - 9 + 3x + 21

$$-2x^2 \; + \; 4x \; + \; 3x \; - \; 9 \; + \; 21$$
$$-2x^2 \; + \; 7x \; - \; 9 \; + \; 21$$
$$-2x^2 \; + \; 7x \; + \; 12$$

---

12 + (-2x) - 3 + 4x

$$-2x \; + \; 4x \; + \; 12 \; - \; 3$$
$$2x \; + \; 12 \; - \; 3$$
$$2x \; + \; 9$$

Combining Like Terms

$$-4 + 9x - 5 - 8x + 1$$

_____

_____

_____

_____

_____

$$5 - 7 + 3x - 4x - 2$$

_____

_____

_____

_____

_____

$$4x - 2x^2 - 9 + 3x + 21$$

_____

_____

_____

_____

_____

$$12 + (-2x) - 3 + 4x$$

_____

_____

_____

_____

_____

Combining Like Terms

$$-8x - 2x^2 + 5 - 5x$$
$$-2x^2 - 8x - 5x + 5$$
$$-2x^2 - 13x + 5$$

$$x^2 - 2x - 9 + 3x - 4x^2 + 4$$
$$x^2 - 4x^2 - 2x + 3x - 9 + 4$$
$$-3x^2 - 2x + 3x - 9 + 4$$
$$-3x^2 + 1x - 9 + 4$$
$$-3x^2 + 1x - 5$$
$$-3x^2 + x - 5$$

$$-9 + 8x + 23 + 7x - 3x$$
$$8x + 7x - 3x - 9 + 23$$
$$12x - 9 + 23$$
$$12x + 14$$

$$5 - (x + 5) - 9x$$
$$-9x - x + 5 - 5$$
$$-10x + 5 - 5$$
$$-10x$$

Combining Like Terms

$-8x - 2x^2 + 5 - 5x$

$x^2 - 2x - 9 + 3x - 4x^2 + 4$

$-9 + 8x + 23 + 7x - 3x$

$5 - (x + 5) - 9x$

Combining Like Terms

$$7x - (3x - 5) + 12$$
$$7x - 3x - 5 + 12$$
$$4x - 5 + 12$$
$$4x + 7$$

$$x - (-3x + 2x^2 - 3) + 9x$$
$$-2x^2 + x + 3x + 9x + 3$$
$$-2x^2 + 13x + 3$$

$$-x + 5 - 3x - 12 + 5x$$
$$-x - 3x + 5x + 5 - 12$$
$$1x + 5 - 12$$
$$1x - 7$$
$$x - 7$$

$$-(2 - 3x) - (-4x - 8)$$
$$3x + 4x - 2 + 8$$
$$7x - 2 + 8$$
$$7x + 6$$

Combining Like Terms

$$7x - (3x - 5) + 12$$

$$x - (-3x + 2x^2 - 3) + 9x$$

$$-x + 5 - 3x - 12 + 5x$$

$$-(2 - 3x) - (-4x - 8)$$

Combining Like Terms

$$(-x - 7) + (8x - 20) - 4$$

$$-x\ +\ 8x\ -7\ -\ 20\ -\ 4$$
$$7x\ -\ 7\ -\ 20\ -\ 4$$
$$7x\ -\ 31$$

$$12x - (7 + 8x) - 9$$

$$12x\ -\ 8x\ -\ 7\ -\ 9$$
$$4x\ -\ 7\ -\ 9$$
$$4x\ -\ 16$$

$$14 - x + 4 + 12x - (-3x)$$

$$-x\ +\ 12x\ +\ 3x\ +\ 14\ +\ 4$$
$$14x\ +\ 14\ +\ 4$$
$$14x\ +\ 18$$

$$-x^2 + 7 + 3x - 5x^2 + 5x - 9$$

$$-x^2\ -\ 5x^2\ +\ 3x\ +\ 5x\ +\ 7\ -\ 9$$
$$-6x^2\ +\ 3x\ +\ 5x\ +\ 7\ -\ 9$$
$$-6x^2\ +\ 8x\ +\ 7\ -\ 9$$
$$-6x^2\ +\ 8x\ -\ 2$$

Combining Like Terms

$$(-x - 7) + (8x - 20) - 4$$

_____
_____
_____
_____
_____

$$12x - (7 + 8x) - 9$$

_____
_____
_____
_____
_____

$$14 - x + 4 + 12x - (-3x)$$

_____
_____
_____
_____
_____

$$-x^2 + 7 + 3x - 5x^2 + 5x - 9$$

_____
_____
_____
_____
_____

Combining Like Terms

$$7 - x - 9x^2 + (3 - 2x)$$
$$9x^2 - x - 2x + 7 + 3$$
$$9x^2 - 3x + 7 + 3$$
$$9x^2 - 3x + 10$$

$$-(-3x - 12) - (x + 7) - 8$$
$$3x - x + 12 - 7 - 8$$
$$2x + 12 - 7 - 8$$
$$2x - 3$$

$$-4x^2 - x + x^2 - 2 + 5x$$
$$-4x^2 + x^2 - x + 5x - 2$$
$$-3x^2 - x + 5x - 2$$
$$-3x^2 + 4x - 2$$

$$3x - x + 9 - 7x + 12 - 11x^2$$
$$-11x^2 + 3x - x - 7x + 9 + 12$$
$$-11x^2 - 5x + 9 + 12$$
$$-11x^2 - 5x + 21$$

Combining Like Terms

$$7 - x - 9x^2 + (3 - 2x)$$

$$-(-3x - 12) - (x + 7) - 8$$

$$-4x^2 - x + x^2 - 2 + 5x$$

$$3x - x + 9 - 7x + 12 - 11x^2$$

Algebra Definitions

Expressions
A mathematical phrase that may contain numbers, variables, or operations, but does not include a relationship symbol such as =, <, or >.

Variables
A letter used that represents a quantity that can change.

Coefficient
The number part of a term with a variable.

Constants
A term containing only numbers. Constants do not include variables.

Binomial
A polynomial with two terms.

Polynomial
An expression consisting of variables and coefficients that involves only the operations of addition, subtraction, multiplication, and non-negative integer exponents.

Inequality
A sentence that states one expression is greater than, less than or equal to another expression

Did you know......

If you add up all the numbers from 1 to 100 consecutively (1 + 2 + 3...) it totals 5050.

A number is divisible by 3 if the sum of its digits is divisible by 3.

If you divide 1 by a number, and divide 1 by that number, you end up with the number you started with.

The opposite sides of a die always add up to 7.

A 'Jiffy' is 1/100th of a second.

Different names for the number 0 are nought, naught, nil, zilch, and zip.

After million, billion, and trillion, there are quadrillion, quintillion, sextillion, septillion, octillion, nonillion, decillion, and undecillion.

The equals sign (=) was invented in 1557 by a mathematician named Robert Recorde.

Combining Like Terms

7y - 2x + (y - 4x) - 12 - 5y

$-2x - 4x + 7y + y - 5y - 12$

$-6x + 7y + y - 5y - 12$

$-6x + 3y - 12$

7a + (x - 2) - 3a + 9x - 5

$7a - 3a + x + 9x - 2 - 5$

$4a + x + 9x - 2 - 5$

$4a + 10x - 2 - 5$

$4a + 10x - 7$

$3x^2$ + 8b - 9x - (-x + 5b)

$3x^2 + 8b - 5b - 9x + x$

$3x^2 + 3b - 9x + x$

$3x^2 + 3b - 8x$

5y + 2x + 5 - 8y - 3 - 7x

$2x - 7x + 5y - 8y + 5 - 3$

$-5x + 5y - 8y + 5 - 3$

$-5x - 3y + 5 - 3$

$-5x - 3y + 2$

Combining Like Terms

$$7y - 2x + (y - 4x) - 12 - 5y$$

$$7a + (x - 2) - 3a + 9x - 5$$

$$3x^2 + 8b - 9x - (-x + 5b)$$

$$5y + 2x + 5 - 8y - 3 - 7x$$

Combining Like Terms

$3a - 9x + (b - x) - 4a + b$

$3a - 4a + b + b - 9x - x$

$-1a + b + b - 9x - x$

$-1a + 2b - 9x - x$

$-a + 2b - 10x$

$2x + 4y + 7 - 3y + 4x + 5$

$2x + 4x + 4y - 3y + 7 + 5$

$6x + 4y - 3y + 7 + 5$

$6x + 1y + 7 + 5$

$6x + y + 12$

$18 - y + 4x - (-x + 2y) - 5$

$4x + x - y - 2y + 18 - 5$

$5x - y - 2y + 18 - 5$

$5x - 3y + 18 - 5$

$5x - 3y + 13$

$x - (a - 3a - 2x) + 5 - 3x$

$-a + 3a + x + 2x - 3x + 5$

$2a + x + 2x - 3x + 5$

$2a + 5$

Combining Like Terms

$$3a - 9x + (b - x) - 4a + b$$

_____

_____

_____

_____

_____

$$2x + 4y + 7 - 3y + 4x + 5$$

_____

_____

_____

_____

_____

$$18 - y + 4x - (-x + 2y) - 5$$

_____

_____

_____

_____

_____

$$x - (a - 3a - 2x) + 5 - 3x$$

_____

_____

_____

_____

_____

Combining Like Terms

$$7y - 8b + (x - 5b) - (2x - 4y)$$
$$-8b \ -5b \ + \ x \ - \ 2x \ + \ 7y \ + \ 4y$$
$$-13b \ + \ x \ - \ 2x \ + \ 7y \ + \ 4y$$
$$-13b \ - \ 1x \ + \ 7y \ + \ 4y$$
$$-13b \ - \ x \ + \ 11y$$

$$4 + 2x + 8 - x + 5y^2$$
$$5y^2 \ + \ 2x \ - \ x \ + \ 4 \ + \ 8$$
$$5y^2 \ + \ 1x \ + \ 4 \ + \ 8$$
$$5y^2 \ + \ x \ + \ 12$$

$$4y + 3x - 2 - 4y + 7x - 12$$
$$3x \ + \ 7x \ + \ 4y \ - \ 4y \ - \ 2 \ - \ 12$$
$$10x \ + \ 4y \ - \ 4y \ - \ 2 \ - \ 12$$
$$10x \ - \ 2 \ - \ 12$$
$$10x \ - \ 14$$

$$(-7 - 4x) - (-5x + 8) - 5$$
$$-4x \ + \ 5x \ - \ 7 \ - \ 8 \ - \ 5$$
$$1x \ - \ 7 \ - \ 8 \ - \ 5$$
$$x \ - \ 20$$

Combining Like Terms

$$7y - 8b + (x - 5b) - (2x - 4y)$$

_____

_____

_____

_____

_____

$$4 + 2x + 8 - x + 5y^2$$

_____

_____

_____

_____

$$4y + 3x - 2 - 4y + 7x - 12$$

_____

_____

_____

_____

$$(-7 - 4x) - (-5x + 8) - 5$$

_____

_____

_____

_____

Simplify and Solve

$$2x + 5 - x + 2 = 10$$

$$2x - x + 5 + 2 = 10$$
$$x + 5 + 2 = 10$$
$$x + 7 = 10$$
$$x = 10 - 7$$
$$x = 3$$

$$(5 - 3x) - 8 + 4x = 18$$

$$-3x + 4x + 5 - 8 = 18$$
$$x + 5 - 8 = 18$$
$$x - 3 = 18$$
$$x = 18 + 3$$
$$x = 21$$

$$-x - 2 + 4x + 5 - 2x = 14$$

$$-x + 4x - 2x - 2 + 5 = 14$$
$$x - 2 + 5 = 14$$
$$x + 3 = 14$$
$$x = 14 - 3$$
$$x = 11$$

$$4 + (-5 - 3x) + 4x = 22$$

$$-3x + 4x + 4 - 5 = 22$$
$$x + 4 - 5 = 22$$
$$x - 1 = 22$$
$$x = 22 + 1$$
$$x = 23$$

Simplify and Solve

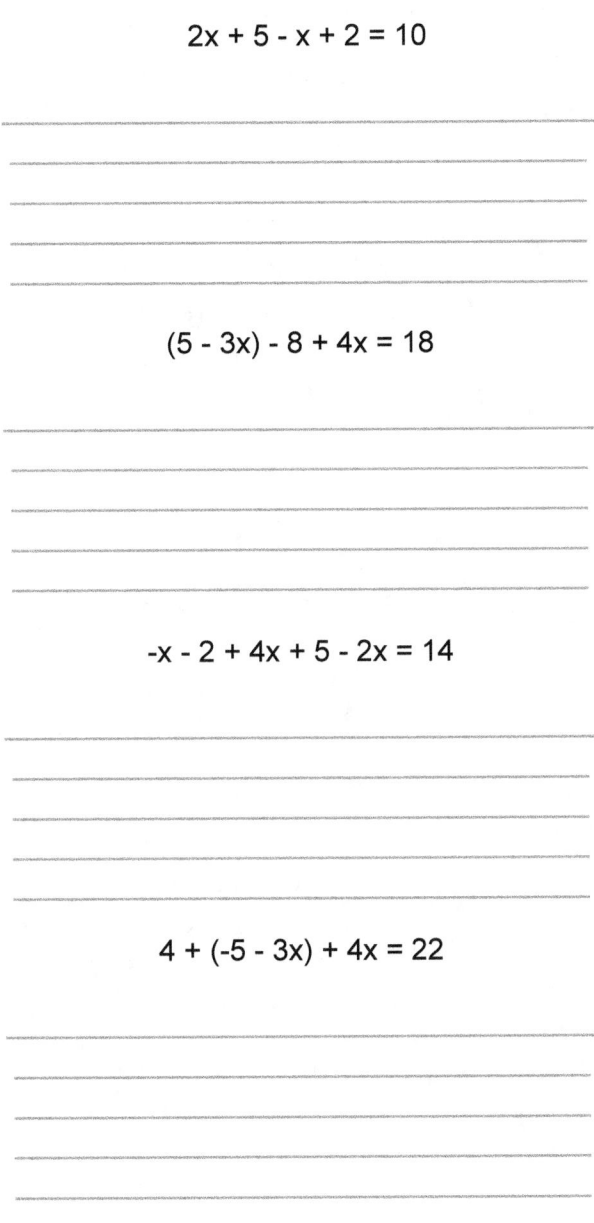

$$2x + 5 - x + 2 = 10$$

$$(5 - 3x) - 8 + 4x = 18$$

$$-x - 2 + 4x + 5 - 2x = 14$$

$$4 + (-5 - 3x) + 4x = 22$$

Simplify and Solve

$$8x - (-2 - 3x) = 14 + 10x$$
$$8x + 3x + 2 = 14 + 10x$$
$$11x + 2 = 14 + 10x$$
$$11x - 10x + 2 = 14$$
$$x = 14 - 2$$
$$x = 12$$

$$5 - 3 + 7x = 9 + 6x - 2$$
$$2 + 7x = 6x + 7$$
$$2 + 7x - 6x = 7$$
$$x = 7 - 2$$
$$x = 5$$

$$-6x - 3 + 2x = 17 - 5x$$
$$-4x - 3 = 17 - 5x$$
$$-4x + 5x - 3 = 17$$
$$x - 3 = 17$$
$$x = 17 + 3$$
$$x = 20$$

$$2 - (-3x - 8) = (2x + 3) + 8$$
$$2 + 3x + 8 = 2x + 11$$
$$3x + 10 = 2x + 11$$
$$3x - 2x + 10 = 11$$
$$x = 11 - 10$$
$$x = 1$$

Simplify and Solve

$$8x - (-2 - 3x) = 14 + 10x$$

$$5 - 3 + 7x = 9 + 6x - 2$$

$$-6x - 3 + 2x = 17 - 5x$$

$$2 - (-3x - 8) = (2x + 3) + 8$$

Simplify and Solve

$$2y - 7 + (2 + 3y) = 4 + 4y$$
$$2y + 3y - 7 + 2 = 4 + 4y$$
$$5y - 7 + 2 = 4 + 4y$$
$$5y - 5 = 4 + 4y$$
$$5y - 4y = 4 + 5$$
$$y = 9$$

$$8 - y + 2 = -4y + 5 + 2y$$
$$-y + 10 = -2y + 5$$
$$-y + 2y + 10 = 5$$
$$y + 10 = 5$$
$$y = 5 - 10$$
$$y = -5$$

$$25 - 15x = 15 - 16x$$
$$25 - 15x + 16x = 15$$
$$-15x + 16x = 15 - 25$$
$$x = 15 - 25$$
$$x = -10$$

$$(4b - 9) - (2b + 7) = 4 + b$$
$$4b - 2b - 9 - 7 = 4 + b$$
$$2b - 9 - 7 = 4 + b$$
$$2b - 16 = 4 + b$$
$$2b - b = 4 + 16$$
$$b = 20$$

Simplify and Solve

$$2y - 7 + (2 + 3y) = 4 + 4y$$

$$8 - y + 2 = -4y + 5 + 2y$$

$$25 - 15x = 15 - 16x$$

$$(4b - 9) - (2b + 7) = 4 + b$$

Simplify and Solve

$$58 = 5x - (12 + 4x) + 14$$
$$58 = 5x - 4x - 12 + 14$$
$$58 = x + 2$$
$$-x = 2 - 58$$
$$-x = -56$$
$$x = 56$$

$$x - 15 = (-5 - 9) - 8$$
$$x - 15 = -22$$
$$x = -22 + 15$$
$$x = -7$$

$$24 = 3x - (-5 + 2x) + 9$$
$$24 = 3x - 2x + 5 + 9$$
$$24 = x + 14$$
$$-x = 14 - 24$$
$$-x = -10$$
$$x = 10$$

$$5 - 4x = x + (2x + 5)$$
$$5 - 4x = 3x + 5$$
$$5 - 4x - 3x = 5$$
$$4x - 3x = 5 - 5$$
$$x = 0$$

Simplify and Solve

$$58 = 5x - (12 + 4x) + 14$$

$$x - 15 = (-5 - 9) - 8$$

$$24 = 3x - (-5 + 2x) + 9$$

$$5 - 4x = x + (2x + 5)$$

Simplify and Solve

$$7 - 8x = -3x - (6x + 5) - 19$$
$$7 - 8x = -9x - 24$$
$$-8x = -9x - 24 - 7$$
$$-8x + 9x = -31$$
$$x = -31$$

$$17x = -(-14x + 4) + 2x$$
$$17x = 14x + 2x - 4$$
$$17x = 16x - 4$$
$$17x - 16x = -4$$
$$1x = -4$$
$$x = -4$$

$$x - 9 = (-2x - 2) - (-2x + 9)$$
$$x - 9 = -2x + 2x - 2 - 9$$
$$x - 9 = -2 - 9$$
$$x = -2 - 9 + 9$$
$$x = -2$$

$$10 - 4 - x = -2x - 22$$
$$6 - x = -2x - 22$$
$$6 - x + 2x = -22$$
$$-x + 2x = -22 - 6$$
$$1x = -28$$
$$x = -28$$

Simplify and Solve

$$7 - 8x = -3x - (6x + 5) - 19$$

$$17x = -(-14x + 4) + 2x$$

$$x - 9 = (-2x - 2) - (-2x + 9)$$

$$10 - 4 - x = -2x - 22$$

Simplify and Solve

$$14x - 3 = (12x - 5) + x$$

$$14x - 3 = 12x + x - 5$$
$$14x - 3 = 13x - 5$$
$$14x - 13x - 3 = -5$$
$$1x = -5 + 3$$
$$x = -2$$

$$4 - x + 2 = 9 - 2x$$

$$-x + 4 + 2 = 9 - 2x$$
$$-x + 2x + 6 = 9$$
$$1x = 9 - 6$$
$$x = 3$$

$$5 + 3x = 2x + 5$$

$$3x = 2x + 5 - 5$$
$$3x - 2x = 5 - 5$$
$$1x = 0$$
$$x = 0$$

$$(-3x + 4) - 2 = 8 - 4x + 1$$

$$-3x + 2 = -4x + 8 + 1$$
$$-3x + 2 = -4x + 9$$
$$-3x = -4x + 9 - 2$$
$$-3x + 4x = 7$$
$$1x = 7$$
$$x = 7$$

Simplify and Solve

$$14x - 3 = (12x - 5) + x$$

_____

_____

_____

_____

_____

$$4 - x + 2 = 9 - 2x$$

_____

_____

_____

_____

_____

$$5 + 3x = 2x + 5$$

_____

_____

_____

_____

_____

$$(-3x + 4) - 2 = 8 - 4x + 1$$

_____

_____

_____

_____

_____

Scientific Notation

$$5.4 \times 10^3$$
$$\underline{\times\ 3.2 \times\ 10^2}$$
$$108$$
$$\underline{+\ 1620}$$
$$17.28 \times 10^5$$
$$1.728 \times 10^6$$

$$4.7 \times 10^3$$
$$\underline{\times\ 2.4 \times\ 10^2}$$
$$188$$
$$\underline{+\ 490}$$
$$6.78 \times 10^5$$

$$7.2 \times 10^4$$
$$\underline{\times\ 5.5 \times\ 10^3}$$
$$360$$
$$\underline{+\ 3600}$$
$$39.60 \times 10^7$$
$$3.96 \times 10^8$$

$$2.3 \times 10^7$$
$$\underline{\times\ 3.7 \times\ 10^4}$$
$$151$$
$$\underline{+\ 690}$$
$$8.41 \times 10^{11}$$

Scientific Notation

$$5.4 \times 10^3$$
$$\underline{\times \ 3.2 \times 10^2}$$

$$4.7 \times 10^3$$
$$\underline{\times \ 2.4 \times 10^2}$$

$$7.2 \times 10^4$$
$$\underline{\times \ 5.5 \times 10^3}$$

$$2.3 \times 10^7$$
$$\underline{\times \ 3.7 \times 10^4}$$

Scientific Notation

$$1.7 \times 10^{-3}$$
$$\underline{\times\ 7.3 \times 10^{2}}$$
$$51$$
$$\underline{+\ 1190}$$
$$12.41 \times 10^{-1}$$
$$1.241 \times 10^{0}$$

$$9.7 \times 10^{-5}$$
$$\underline{\times\ 7.3 \times 10^{-2}}$$
$$291$$
$$\underline{+\ 6790}$$
$$70.81 \times 10^{-7}$$
$$7.081 \times 10^{-6}$$

$$8.8 \times 10^{-11}$$
$$\underline{\times\ 5.9 \times 10^{3}}$$
$$792$$
$$\underline{+\ 4400}$$
$$51.92 \times 10^{-8}$$
$$5.192 \times 10^{-7}$$

$$2.7 \times 10^{7}$$
$$\underline{\times\ 1.9 \times 10^{-23}}$$
$$243$$
$$\underline{+\ 270}$$
$$5.13 \times 10^{-16}$$

Scientific Notation

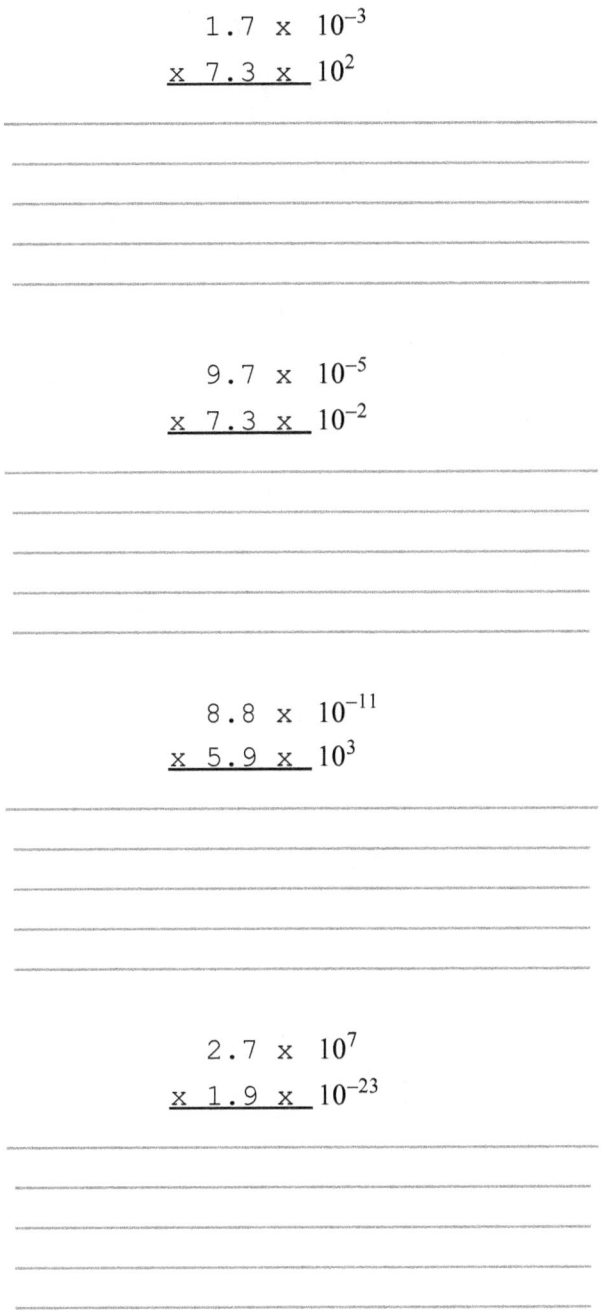

$$1.7 \times 10^{-3}$$
$$\times\ 7.3 \times 10^{2}$$

$$9.7 \times 10^{-5}$$
$$\times\ 7.3 \times 10^{-2}$$

$$8.8 \times 10^{-11}$$
$$\times\ 5.9 \times 10^{3}$$

$$2.7 \times 10^{7}$$
$$\times\ 1.9 \times 10^{-23}$$

Scientific Notation

$$8.65 \times 10^{-9}$$
$$\underline{\times\ 9.5 \times 10^{2}}$$
$$4325$$
$$\underline{+\ 77850}$$
$$82.175 \times 10^{-7}$$
$$8.2175 \times 10^{-6}$$

$$5.59 \times 10^{-32}$$
$$\underline{\times\ 9.4 \times 10^{-21}}$$
$$2236$$
$$\underline{+\ 50310}$$
$$52.546 \times 10^{-53}$$
$$5.2546 \times 10^{-52}$$

$$1.99 \times 10^{-5}$$
$$\underline{\times\ 8.7 \times 10^{34}}$$
$$1393$$
$$\underline{+\ 15920}$$
$$17.313 \times 10^{29}$$
$$1.7313 \times 10^{30}$$

$$6.7 \times 10^{-9}$$
$$\underline{\times\ 1.1 \times 10^{40}}$$
$$67$$
$$\underline{+\ 670}$$
$$7.37 \times 10^{31}$$

Scientific Notation

$$8.65 \times 10^{-9}$$
$$\underline{\times\ 9.5 \times 10^{2}}$$

$$5.59 \times 10^{-32}$$
$$\underline{\times\ 9.4 \times 10^{-21}}$$

$$1.99 \times 10^{-5}$$
$$\underline{\times\ 8.7 \times 10^{34}}$$

$$6.7 \times 10^{-9}$$
$$\underline{\times\ 1.1 \times 10^{40}}$$

Scientific Notation

$$2.3 \times 10^3$$
$$\underline{+\ 5.2 \times 10^2}$$

$$2.30 \times 10^3$$
move decimal $$\underline{+\ 0.52 \times 10^3}$$
$$2.82 \times 10^3$$

$$4.1 \times 10^5$$
$$\underline{+\ 5.7 \times 10^2}$$

$$4.1000 \times 10^5$$
move decimal $$\underline{+\ 0.0057 \times 10^5}$$
$$4.1057 \times 10^5$$

$$7.2 \times 10^9$$
$$\underline{+\ 8.7 \times 10^5}$$

$$7.20000 \times 10^9$$
move decimal $$\underline{+\ 0.00087 \times 10^9}$$
$$7.20087 \times 10^9$$

Scientific Notation

$$2.3 \times 10^3$$
$$+ \; 5.2 \times 10^2$$

$$4.1 \times 10^5$$
$$+ \; 5.7 \times 10^2$$

$$7.2 \times 10^9$$
$$+ \; 8.7 \times 10^5$$

Scientific Notation

$$8.12 \times 10^8$$
$$\underline{+ \quad 7.8 \times 10^6}$$

$$8.12 \quad \times \ 10^8$$
move decimal $$\underline{+ \ 0.078 \times \ 10^8}$$
$$8.198 \times \ 10^8$$

$$7.9 \times \ 10^{-7}$$
$$\underline{+ \ 4.79 \times \ 10^{-4}}$$

move decimal $$0.0079 \times \ 10^{-4}$$
$$\underline{+ \ 4.7900 \times \ 10^{-4}}$$
$$4.7979 \times \ 10^{-4}$$

$$2.65 \times \ 10^{-8}$$
$$\underline{+ \ 9.33 \times \ 10^{-7}}$$

move decimal $$0.265 \times \ 10^{-7}$$
$$\underline{+ \ 9.330 \times \ 10^{-7}}$$
$$9.595 \times \ 10^{-7}$$

Scientific Notation

$$8.12 \times 10^8$$
$$+ \quad 7.8 \times 10^6$$

$$7.9 \times 10^{-7}$$
$$+ \ 4.79 \times 10^{-4}$$

$$2.65 \times 10^{-8}$$
$$+ \ 9.33 \times 10^{-7}$$

Scientific Notation

$$5.002 \times 10^5$$
$$+ \quad 2.5 \times 10^6$$

move decimal $\quad 0.5002 \quad \times \ 10^6$
$$+ \quad 2.5000 \quad \times \ 10^6$$
$$3.0002 \quad \times \ 10^6$$

$$9.12 \times 10^{-3}$$
$$+ \ 2.04 \times 10^{-4}$$

$$9.120 \times 10^{-3}$$
move decimal $\quad + \ 0.204 \times \ 10^{-3}$
$$9.324 \times 10^{-3}$$

$$8.99 \times 10^{-11}$$
$$+ \ 9.79 \times \ 10^{-9}$$

move decimal $\quad 0.0899 \times 10^{-9}$
$$+ \ 9.7900 \times \ 10^{-9}$$
$$9.8799 \times 10^{-9}$$

Scientific Notation

$$5.002 \times 10^5$$
$$+ \quad 2.5 \times 10^6$$

$$9.12 \times 10^{-3}$$
$$+ 2.04 \times 10^{-4}$$

$$8.99 \times 10^{-11}$$
$$+ 9.79 \times 10^{-9}$$

Scientific Notation

$$4.99 \times 10^8$$
$$-\underline{3.8 \times 10^6}$$

$$4.990 \times 10^8$$
move decimal $\quad -\underline{0.038 \times10^8}$
$$4.952 \times 10^8$$

$$7.08 \times 10^{-5}$$
$$-\underline{4.42 \times 10^{-6}}$$

$$7.080 \times 10^{-5}$$
move decimal $\quad -\underline{0.442 \times10^{-5}}$
$$6.638 \times 10^{-5}$$

$$8.55 \times 10^{-1}$$
$$-\underline{1.24 \times 10^{1}}$$

move decimal $\quad 0.0855 \times 10^{1}$
$$-\underline{1.2400 \times10^{1}}$$
$$-1.1545 \times 10^{1}$$

Scientific Notation

$$4.99 \times 10^8$$
$$- \quad 3.8 \times 10^6$$

$$7.08 \times 10^{-5}$$
$$- \ 4.42 \times 10^{-6}$$

$$8.55 \times 10^{-1}$$
$$- \ 1.24 \times 10^1$$

Scientific Notation

$$1.09 \times 10^{-1}$$
$$- \underline{\ 4.8 \times\ } 10^{-2}$$

$$1.09 \times 10^{-1}$$
$move\ decimal \quad - \underline{\ 0.48 \times\ } 10^{-1}$
$$0.61 \times 10^{-1}$$
$$6.1 \quad \times\ 10^{-2}$$

$$3.27 \times 10^{4}$$
$$- \underline{\ 8.29 \times\ } 10^{5}$$

$move\ decimal \quad 0.327 \times 10^{5}$
$$- \underline{\ 8.290 \times\ } 10^{5}$$
$$-7.963 \times 10^{5}$$

$$7.57 \times 10^{12}$$
$$- \underline{\ 5.98 \times\ } 10^{10}$$

$$7.5700 \times 10^{12}$$
$move\ decimal \quad - \underline{\ .0598 \times\ } 10^{12}$
$$7.5102 \times 10^{12}$$

Scientific Notation

$$1.09 \times 10^{-1}$$
$$- \underline{4.8 \times} 10^{-2}$$

$$3.27 \times 10^{4}$$
$$- \underline{8.29 \times} 10^{5}$$

$$7.57 \times 10^{12}$$
$$- \underline{5.98 \times} 10^{10}$$

Scientific Notation

$$4.57 \times 10^{-23}$$
$$- \ \underline{2.1 \times 10^{-21}}$$

move decimal
$$0.0457 \times 10^{-21}$$
$$- \ \underline{2.1000 \times 10^{-21}}$$
$$-2.0543 \times 10^{-21}$$

$$2.11 \times 10^{9}$$
$$- \ \underline{7.202 \times 10^{9}}$$

$$2.110 \times 10^{9}$$
$$- \ \underline{7.202 \times 10^{9}}$$
$$-5.092 \times 10^{9}$$

$$8.55 \times 10^{-3}$$
$$- \ \underline{1.24 \times 10^{-2}}$$

move decimal
$$0.855 \times 10^{-2}$$
$$- \ \underline{1.240 \times 10^{-2}}$$
$$-0.385 \times 10^{-2}$$
$$-3.85 \times 10^{-3}$$

Scientific Notation

$$4.57 \times 10^{-23}$$
$$-\ \underline{2.1 \times} \ 10^{-21}$$

$$2.11 \times 10^{9}$$
$$-\ \underline{7.202 \times} \ 10^{9}$$

$$8.55 \times 10^{-3}$$
$$-\ \underline{1.24 \times} \ 10^{-2}$$

Working with negative numbers

Add/Subtract

Like signs - Add	5 $\underline{+3}$ 8	-5 $\underline{+-3}$ -8
Unlike signs - Subtract	5 $\underline{-3}$ 2	-5 $\underline{+3}$ -2

Multiply/Divide

Like signs - Positive	5 $\underline{\text{x } 3}$ 15	-5 $\underline{\text{x } -3}$ 15
Unlike signs - Negative	5 $\underline{\text{x} -3}$ -15	-5 $\underline{\text{x } 3}$ -15

Algebra rules for arithmetic

$$a(b + c) = ab + ac$$

$$a\left(\tfrac{b}{c}\right) = \tfrac{ab}{c}$$

$$\tfrac{a}{b} + \tfrac{c}{d} = \tfrac{ad + bc}{bd}$$

$$\tfrac{a + b}{c} = \tfrac{a}{c} + \tfrac{b}{x}$$

$$\tfrac{ac + bc}{c} = a + b$$

Converting to Scientific Notation

Place decimal behind leading digit and count how many digits are behind the leading digit for the power of 10.

12,000	1.2×10^{4}
120,000	1.2×10^{5}

For decimals, move decimal behind leading numerical digit and count how many digits it was moved for the negative exponent of 10.

.00012	1.2×10^{-4}
.000012	1.2×10^{-5}

Other books from Timothy Schablin Mathematics

equals(me)
Pre-Algebra Practice

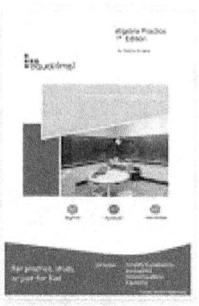

equals(me)
Algebra Practice

equals(me)
Radicals Practice

Fun with PEMDAS
PEMDAS Practice

Available at Amazon Books or

https://timothyschablin.wixsite.com/equalsme

About the author

Timothy Schablin is a graduate of the Hutchinson Technical College where he studied algebra, trigonometry, physics, mathematical techniques, and technical related fields. He also holds two certificates of physics from Davidson College, AP Physics I & AP Physics II: Challenging Concepts.

Timothy Schablin tutors math to 5th, 6th, 7th, and 8th graders at a local middle school. He is also a member of Minnesota MathCorps and has authored mathematical software.

Besides studying & tutoring math and physics, Timothy enjoys astronomy. He spends vacation time canoeing the Minnesota River bottom.